Easily Master Trigonometry

TABLE OF CONTENTS

Introduction

1. Chapter-1: The Triangle

    1.1 Types of Triangle

    1.2 Perimeter and Area of Triangle

    1.3 Trigonometry of a Right-angled Triangle

2. Chapter-2: The Unit Circle

    2.1 Degree and Radian

    2.2 Conversion between Degree and Radian

3. Chapter-3: Trigonometric Functions

    3.1 Trigonometric Functions with respect to a unit circle

    3.2 Trigonometric Table

    3.3 Concepts of Signs in various Quadrants

    3.4 Trigonometric function values of non-standard angles.

    3.5 Trigonometric Identities

4. Chapter:4- Heights & Distances

INTRODUCTION

The term "Trigonometry" is derived from the Greek terms "trigonon" which means triangle and "metron" which means measure.

In ancient times it was mainly developed to calculate sides and angles of triangle or shapes in which triangles were embedded. Later it found extensive use in Astrology to calculate distances of earth from various planetary bodies using the shadow casted on a body at a particular angle.

In India, the concepts of trigonometry were used in building "Jantar Mantar", built by Maharaja Jai Singh II of Jaipur in the year 1724 to predict planetary movements

CHAPTER –1: THE TRIANGLE

WHAT IS A TRIANGLE?

A triangle is a 2-dimensional, 3-sided polygon with 3 vertices and 3 edges with the sum of all 3 internal angles being equal to 180°

TYPES OF TRIANGLE:

There are two classification of Triangles, one is based on lengths of sides and the other is based on measure of internal angles.

1. Based on lengths of sides:

    Scalene Triangle: A scalene triangle is a type of triangle, in which all the three sides have different lengths. Due to this, the three angles are also different from each other.

    Isosceles Triangle: In an isosceles triangle, two sides have equal length. The two angles opposite to the two equal sides also equal to each other.

    Equilateral Triangle : An equilateral triangle has all three sides equal to each other. Due to this all the internal angles are of equal degrees, i.e. each of the angles is 60°

2. Based on internal angle measurement:

Acute Angled Triangle: An acute triangle has all of its angles less than 90°.

Right-angled Triangle: In a right triangle, one of the angles isequal to 90° or right angle.

Obtuse-angles Triangle: An obtuse angle has any of its internal angle as more than 90°

PERIMETER AND AREA OF A TRIANGLE:

Perimeter: It is the sum of all the sides of the triangle.

Area : It is the region occupied by the triangle in the 2-D plane.

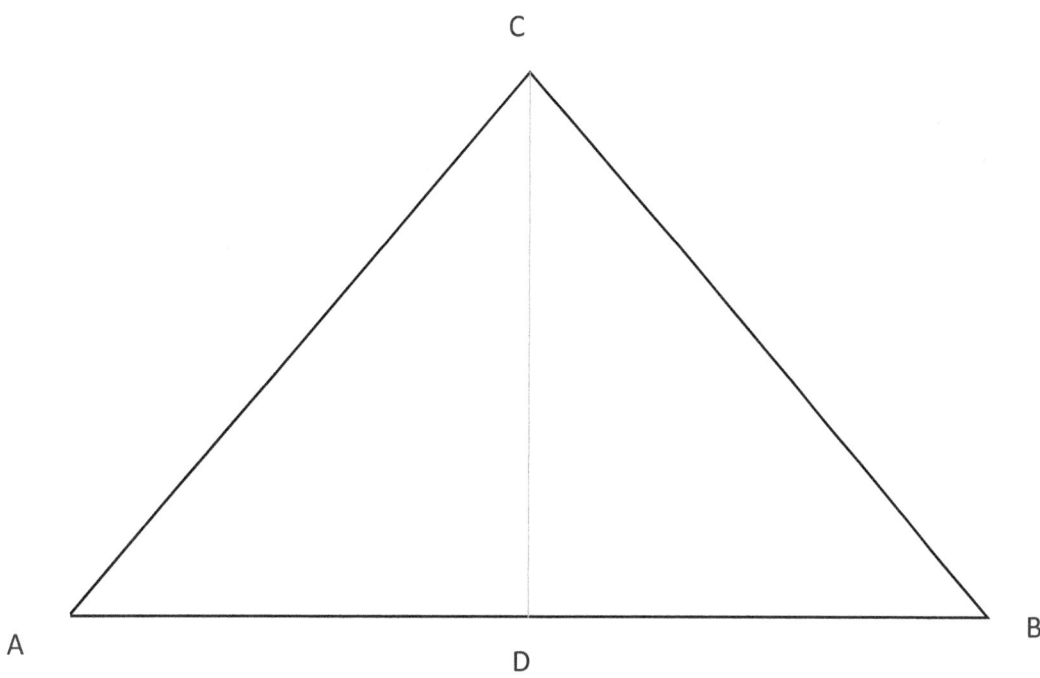

Perimeter = AB + BC + CA

Area of the Triangle = $\frac{1}{2} \cdot Base \cdot Height$

$= \frac{1}{2} \cdot AB \cdot CD$

Area of the triangle using Heron's Formula

Let the sides of the triangle be measured as a, b & c and s be the semi-perimeter, then;

$$s = \frac{(a+b+c)}{2}$$

$$Area = \sqrt{s \cdot (s-a)(s-b)(s-c)}$$

Question: Let ABC be a triangle with side lengths AB = 3 units, BC = 4 units & CA = 5 units respectively. Find the are of the triangle.

Solution:

By Heron's Formula, we have;

$$s = \frac{(3+4+5)}{2}$$

$$s = 6$$

$$Area = \sqrt{6 \cdot (6-3) \cdot (6-4) \cdot (6-5)}$$

$$Area = \sqrt{6 \cdot 6}$$

$$Area = 6 \: squnits$$

Alternatively, the triangle with the given side measurements follows the properties of a right-angled triangle, mainly Pythagoras Theorem.

From Pythagoras Theorem we have;

$$5^2 = 3^2 + 4^2$$

$$25 = 9 + 16$$

Therefore we can say that side measuring 5 units is hypotenuse and the rest are base and altitude depending upon frame of reference.

So,

$$Area = \frac{1}{2} \cdot 3 \cdot 4$$

$$Area = 6 \; sq \; units$$

TRIGONOMETRIC RATIOS WITH RESPECT TO A TRIANGLE

Let's have a look at the trigonometric ratios with respect to a right-angled triangle, right-angled at B.

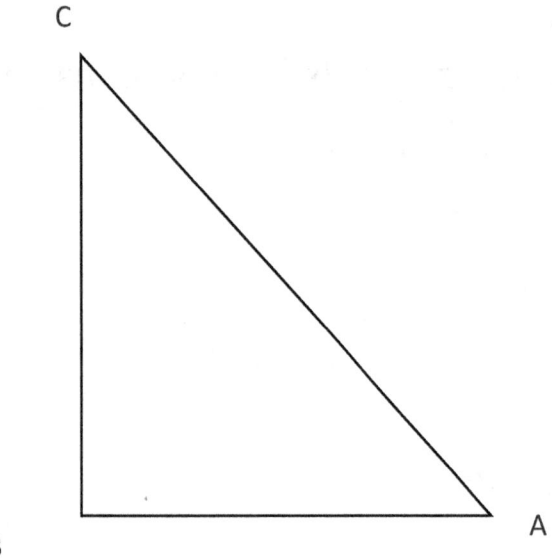

With respect to the △□□□, the trigonometric ratios are as follows:

Sin(A) = Altitude/Hypotenuse

  = BC/AC

Cos(A) = Base/Hypotenuse

  = AB/AC

Tan(A) = Altitude/Base

  = BC/AB

Question: In a right-angled triangle ABC, right angled at B, the base is of length 3 units and altitude is of length 4 units. Find the trigonometric ratios.

Solution: Let's draw the figure for easier understanding;

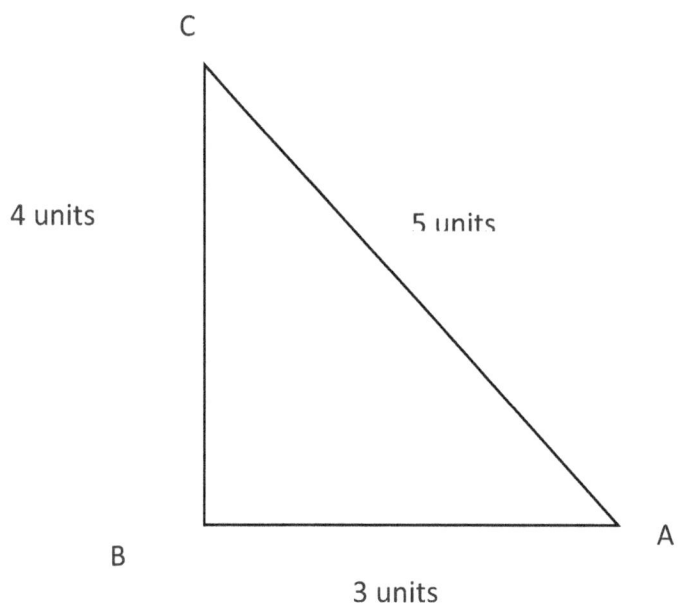

Applying Pythagoras theorem,

$$AC^2 = AB^2 + BC^2$$

$$AC^2 = 3^2 + 4^2$$

$$AC = \sqrt{25} = 5\ units$$

SinA = BC/AC

SinA = 4/5 ,

CosA = AB/AC

CosA = 3/5,

TanA = BC/AB

TanA = 4/3,

CosecA = 1/SinA

CosecA = 5/4,

SecA = 1/CosA

SecA = 5/3,

CotA = 1/TanA

CotA = 3/4

CHAPTER-2: THE UNIT CIRCLE

Degree & Radian

Before starting with trigonometry, we must understand the unit in which an angle is measured. For this, basic understanding of the circle is most important. Refer the unit circle given below:

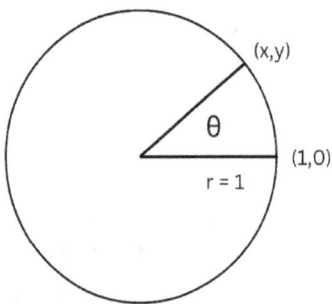

The circumference of the circle is given by $2\pi r$, which here, as this is a unit circle comes to 2π.

Now the angle θ can be measured in two ways:

1. Degree: The complete circle is 360 degrees, if the circle is divided into 360 equal parts, then each such arc shall measure 1 degree of θ.

2. Radian: Here the unit circle has a circumference of 2π, if the circle is again divided into 360 equal parts, then each such arc shall measure to $\frac{2\pi}{360}$ radian.

Conversion:

To understand the conversion better, let's think of a quarter of the

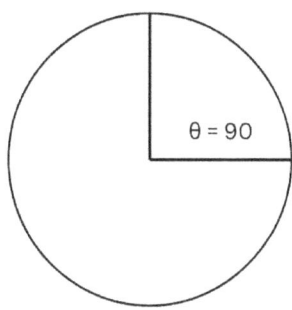

above circle as drawn below:

Here, the quarter of the circle subtends an angle of 90 in the center, while length of the arc shall be $2\pi/4$ which is equal to $\pi/2$.

Hence it can be said that,

$$90° = \frac{\pi}{2} \text{ Radians}$$

$$1° = \frac{\pi}{180} \text{ Radians}$$

OR

π Radians = 180°

Exercise:

Convert the following angles in degrees to radians:

    1. 150°   2. 270°   3. 120°

Solution :

1. $150° \cdot \frac{\pi}{180}$ radians

   $= \frac{5\pi}{6}$ radians

2. $270° \cdot \frac{\pi}{180}$ radians

   $= \frac{3\pi}{2}$ radians

3. $120° \cdot \frac{\pi}{180}$ radians

   $= \frac{2\pi}{3}$

Convert the following angles from radians to degrees:

1.) $\frac{\pi}{2}$ 2.) 2π 3.) $\frac{\pi}{4}$

Solution:

1. $\frac{\pi}{2} \cdot \frac{180°}{\pi}$

   $= 90°$

2. $2\pi \cdot \frac{180}{\pi}$

   $= 360°$

3. $\frac{\pi}{4} \cdot \frac{180}{\pi}$

   $= 45°$

CHAPTER–3: TRIGONOMETRIC FUNCTIONS

Trigonometric Functions

There are 6 basic trigonometric functions namely: sine, cosine, tangent, cosecant, secant & cotangent which are denoted respectively as sin, cos, tan, csc, sec & cot.

Each of these functions represents a ratio of lengths which we will soon find out.

Refer the figure given below:

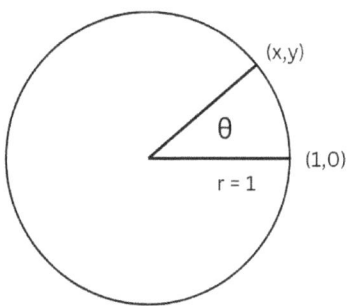

Here, the horizontal distance from center of the circle, x is given by x = radius of the circle * cos(θ), since here we have a unit circle, therefore x = cos(θ).

Similarly, the vertical distance of the point denoted by Cartesian coordinates (x,y) from the center of the circle, y is given by y = radius of the circle * sin(θ), since here we have a unit circle, therefore y = sin(θ).

$$tan\theta = \frac{sin\theta}{cos\theta}$$

$$= \frac{y}{x}$$

$$sec\theta = \frac{1}{cos\theta}$$

$$= \frac{1}{x}$$

$$csc\theta = \frac{1}{sin\theta}$$

$$= \frac{1}{y}$$

$$cot\theta = \frac{1}{tan\theta}$$

$$= \frac{x}{y}$$

Trigonometric Table

The reader must remember the values of trigonometric functions of certain standard angles, the table is given below:

$\dfrac{\theta}{Function}$	0°	30°	45°	60°	90°
$\sin\theta$	0	$\dfrac{1}{2}$	$\dfrac{1}{\sqrt{2}}$	$\dfrac{\sqrt{3}}{2}$	1
$\cos\theta$	1	$\dfrac{\sqrt{3}}{2}$	$\dfrac{1}{\sqrt{2}}$	$\dfrac{1}{2}$	0
$\tan\theta$	0	$\dfrac{1}{\sqrt{3}}$	1	$\sqrt{3}$	∞
$\csc\theta$	∞	2	$\sqrt{2}$	$\dfrac{2}{\sqrt{3}}$	1
$\sec\theta$	1	$\dfrac{2}{\sqrt{3}}$	$\sqrt{2}$	2	∞
$\cot\theta$	∞	$\sqrt{3}$	1	$\dfrac{1}{\sqrt{3}}$	0

Now that we are familiar with the trigonometric function values of certain standard angles, we would learn the concept of finding the values of the functions at other angles and their signs.

Concept of signs with respect to quadrants of a circle.

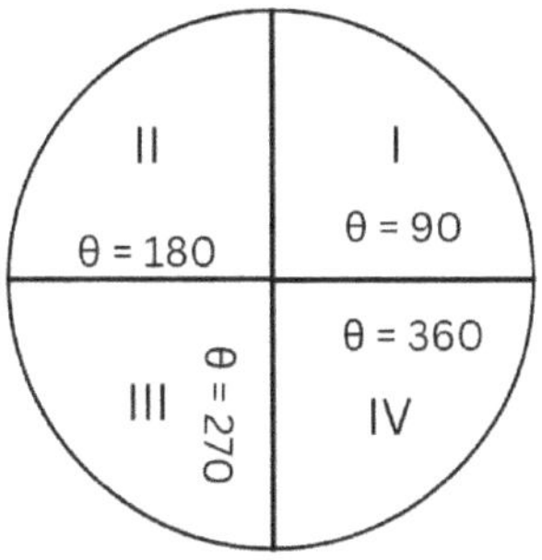

Refer the above figure, the four quadrants are marked as Q-I, Q-II, Q-III, Q-IV where;

- Q-I   : $0 \leq \theta \leq 90$
- Q-II : $90 \leq \theta \leq 180$
- Q-III : $180 \leq \theta \leq 270$
- Q-IV : $270 \leq \theta \leq 360$

Concept of ALL, SIN, TAN, COS

Here the discussion will be with respect to sin, cos and tan.

In Q-I, **ALL** are positive ie; the values of sin, cos and tan at any given angle in the range of $0 \leq \theta \leq 90$, the values of all the functions shall be positive.

In Q-II, only **SIN** is positive ie; the values of sin at any given angle in the range of $90 \leq \theta \leq 180$, the values of only **SIN** shall be positive.

In Q-III, only **TAN** is positive ie; the values of tan at any given angle in the range of $180 \leq \theta \leq 270$, the values of only **TAN** shall be positive. In Q-IV, only **COS** are positive ie; the values of cos at any given angle in the range of $270 \leq \theta \leq 360$, the values of only **COS** shall be positive.

It is worthwhile to mention here the significance of direction of angle based on frame of reference.

For example; an angle of 150 degree measured anticlockwise from 0 degree line shall lie in the second quadrant, the same angle can also be written as -210 degree if it is measured clockwise from 0 degree line, here the negative sign occurs as by standard any clockwise measurement is taken as negative while any anticlockwise measurement is taken as positive.

θ	sin θ	cos θ	tan θ	cot θ	sec θ	cosec θ
-θ	-sinθ	cosθ	-tanθ	-cotθ	secθ	-cosecθ
90-θ	cosθ	sinθ	cotθ	tanθ	cosecθ	secθ
90+θ	cosθ	-sinθ	-cotθ	-tanθ	-cosecθ	secθ
180-θ	sinθ	-cosθ	-tanθ	-cotθ	-secθ	cosecθ
180+θ	-sinθ	-cosθ	tanθ	cotθ	-secθ	-cosecθ
270-θ	-cosθ	-sinθ	cotθ	tanθ	-cosecθ	-secθ
270+θ	-cosθ	sinθ	-cotθ	-tanθ	cosecθ	-secθ
360-θ	-sinθ	cosθ	-tanθ	-cotθ	secθ	-cosecθ
360+θ	sinθ	cosθ	tanθ	cotθ	secθ	cosecθ

Exercise: Use the concepts learned in the previous pages to imagine the position of the angle and then determine the values given in the above table. For consideration, imagine ☐ to be in the 1st quadrant.

Trigonometric function values of non-standard angles

1. Find the value of $\sin(150°)$

Sol:

$\sin(150°) = \sin(90° + 60°)$

$\qquad\qquad = \cos(60°)$ {using identity given in the previous table}

$\qquad\qquad = 1/2$

2. Find the value of $\cos(-210°)$

Sol:

$\cos(-210°) = \cos(210°)$ {using $\cos(-θ) = \cos(θ)$ }

$\cos(210°) = \cos(180° + 30°)$

$\qquad\qquad = -\cos(30°)$

$\qquad\qquad = -\frac{\sqrt{3}}{2}$

PRACTICE PROBLEMS

P-1 Conversion of angles from degrees to radians.

1. Convert 30 degrees to radians.

2. How many radians are there in 90 degrees?

3. Find the radian measure of 45 degrees.

4. What is the radian equivalent of 180 degrees?

5. Convert 60 degrees to radians.

6. How many radians is 120 degrees?

7. What is the radian measure of 270 degrees?

8. Convert 360 degrees to radians.

9. Find the radian measure of 15 degrees.

10. How many radians are there in 75 degrees?

11. What is the radian equivalent of 225 degrees?

12. Convert 150 degrees to radians.

13. How many radians is 240 degrees?

14. Find the radian measure of 135 degrees.

15. Convert 300 degrees to radians.

Solutions:

1. Convert 30 degrees to radians.

$$30° \cdot \frac{\pi}{180°} = \frac{\pi}{6}$$

2. How many radians are there in 90 degrees?

$$90° \cdot \frac{\pi}{180°} = \frac{\pi}{2} \quad \text{radians.}$$

3. Find the radian measure of 45 degrees.

$$45° \cdot \frac{\pi}{180°} = \frac{\pi}{4} \quad \text{radians}$$

4. What is the radian equivalent of 180 degrees?

$$180° \cdot \frac{\pi}{180°} = \pi \quad \text{radians}$$

5. Convert 60 degrees to radians.

$$60° \cdot \frac{\pi}{180°} = \frac{\pi}{3} \quad \text{radians}$$

6. How many radians is 120 degrees?

$$120° \cdot \frac{\pi}{180°} = \frac{2\pi}{3} \quad \text{radians}$$

7. What is the radian measure of 270 degrees?

$$270° \cdot \frac{\pi}{180°} = \frac{3\pi}{2} \quad \text{radians}$$

8. Convert 360 degrees to radians.

$$360° \cdot \frac{\pi}{180°} = 2\pi \qquad \text{radians}$$

9. Find the radian measure of 15 degrees.

$$15° \cdot \frac{\pi}{180°} = \frac{\pi}{12} \qquad \text{radians}$$

10. How many radians are there in 75 degrees?

$$75° \cdot \frac{\pi}{180°} = \frac{5\pi}{12} \qquad \text{radians}$$

11. What is the radian equivalent of 225 degrees?

$$225° \cdot \frac{\pi}{180°} = \frac{5\pi}{4} \qquad \text{radians}$$

12. Convert 150 degrees to radians.

$$150° \cdot \frac{\pi}{180°} = \frac{5\pi}{6} \qquad \text{radians}$$

13. How many radians is 240 degrees?

$$240° \cdot \frac{\pi}{180°} = \frac{4\pi}{3} \qquad \text{radians}$$

14. Find the radian measure of 135 degrees.

$$135° \cdot \frac{\pi}{180°} = \frac{3\pi}{4} \qquad \text{radians}$$

15. Convert 300 degrees to radians.

$$300° \cdot \frac{\pi}{180°} = \frac{5\pi}{3} \qquad \text{radians}$$

P-2 Conversion of angles from radians to degrees.

1. Convert $\frac{\pi}{6}$ radians to degrees.

2. How many degrees are there in $\frac{\pi}{3}$ radians?

3. Find the degree measure of $\frac{\pi}{4}$ radians.

4. What is the degree equivalent of $\frac{\pi}{2}$ radians?

5. Convert π radians to degrees.

6. How many degrees is $\frac{3\pi}{2}$ radians?

7. What is the degree measure of 2π radians?

8. Convert $\frac{2\pi}{3}$ radians to degrees.

9. Find the degree measure of $\frac{5\pi}{6}$ radians.

10. How many degrees are there in $\frac{7\pi}{6}$ radians?

11. What is the degree equivalent of $\frac{5\pi}{4}$ radians?

12. Convert $\frac{3\pi}{4}$ radians to degrees.

13. How many degrees is $\frac{4\pi}{3}$ radians?

14. Find the degree measure of $\frac{5\pi}{3}$ radians.

15. Convert $\frac{7\pi}{4}$ radians to degrees.

Solutions:

Convert $\frac{\pi}{6}$ radians to degrees.

$$\frac{\pi}{6} \cdot \frac{180}{\pi} = 30°$$

How many degrees are there in $\frac{\pi}{3}$ radians?

$$\frac{\pi}{3} \cdot \frac{180}{\pi} = 60°$$

Find the degree measure of $\frac{\pi}{4}$ radians

$$\frac{\pi}{4} \cdot \frac{180}{\pi} = 45°$$

What is the degree equivalent of $\frac{\pi}{2}$ radians?

$$\frac{\pi}{2} \cdot \frac{180}{\pi} = 90°$$

Convert π radians to degrees.

$$\pi \cdot \frac{180}{\pi} = 180°$$

How many degrees is $\frac{3\pi}{2}$ radians?

$$\frac{3\pi}{2} \cdot \frac{180}{\pi} = 270°$$

What is the degree measure of 2π radians?

$$2\pi \cdot \frac{180}{\pi} = 360°$$

Convert $\frac{2\pi}{3}$ radians to degrees.

$$\frac{2\pi}{3} \cdot \frac{180}{\pi} = 120°$$

Find the degree measure of $\frac{5\pi}{6}$ radians.

$\frac{5\pi}{6} \cdot \frac{180}{\pi} = 150°$

How many degrees are there in $\frac{7\pi}{6}$ radians?

$$\frac{7\pi}{6} \cdot \frac{180}{\pi} = 210°$$

What is the degree equivalent of $\frac{5\pi}{4}$ radians?

$\frac{5\pi}{4} \cdot \frac{180}{\pi} = 225°$

Convert $\frac{3\pi}{4}$ radians to degrees.

$$\frac{3\pi}{4} \cdot \frac{180}{\pi} = 135°$$

How many degrees is $\frac{4\pi}{3}$ radians?

$\frac{4\pi}{3} \cdot \frac{180}{\pi} = 240°$

Find the degree measure of $\frac{5\pi}{3}$ radians.

$$\frac{5\pi}{3} \cdot \frac{180}{\pi} = 300°$$

Convert 7π/4 radians to degrees.

$$\frac{7\pi}{4} \cdot \frac{180}{\pi} = 315°$$

Trigonometric Identities

1. Pythagorean Identities:

$$\sin^2(\theta) + \cos^2(\theta) = 1$$

$$\tan^2(\theta) + 1 = \sec^2(\theta)$$

$$1 + \cot^2(\theta) = \csc^2(\theta)$$

2. Reciprocal Identities:

$$\csc(\theta) = \frac{1}{\sin(\theta)}$$

$$\sec(\theta) = \frac{1}{\cos(\theta)}$$

$$\cot(\theta) = \frac{1}{\tan(\theta)}$$

3. Quotient Identities:

$$\tan(\theta) = \frac{\sin(\theta)}{\cos(\theta)}$$

$$\cot(\theta) = \frac{\cos(\theta)}{\sin(\theta)}$$

4. Co-Function Identities:

$\sin(\theta) = \cos\left(\frac{\pi}{2} - \theta\right)$

$\cos(\theta) = \sin\left(\frac{\pi}{2} - \theta\right)$

$\tan(\theta) = \cot\left(\frac{\pi}{2} - \theta\right)$

5. Even-Odd Identities:

$\sin(-\theta) = -\sin(\theta)$

$\cos(-\theta) = \cos(\theta)$

$\tan(-\theta) = -\tan(\theta)$

6. Sum and Difference Identities:

$\sin(A \pm B) = \sin(A)\cos(B) \pm \cos(A)\sin(B)$

$\cos(A \pm B) = \cos(A)\cos(B) \mp \sin(A)\sin(B)$

$\tan(A \pm B) = \dfrac{(\tan(A) \pm \tan(B))}{(1 \mp \tan(A)\tan(B))}$

7. Double Angle Identities:

$\sin(2\theta) = 2\sin(\theta)\cos(\theta)$

$\cos(2\theta) = \cos^2(\theta) - \sin^2(\theta)$

$\tan(2\theta) = \dfrac{2\tan(\theta)}{(1 - \tan^2(\theta))}$

8. Half Angle Identities:

$$\sin\left(\frac{\theta}{2}\right) = \sqrt{\frac{(1-\cos\theta)}{2}}$$

$$\cos\left(\frac{\theta}{2}\right) = \sqrt{\frac{(1+\cos(\theta))}{2}}$$

$$\tan(2\theta) = \frac{2\tan(\theta)}{1-\tan^2(\theta)}$$

Trigonometric values of non-standard angles

1. Find the value of sin(75°):

sin(75°) = sin(45° + 30°)

= sin(45°) * cos(30°) + cos(45°) * sin(30°)

$= \left(\frac{1}{\sqrt{2}} \cdot \frac{\sqrt{3}}{2}\right) + \left(\frac{1}{\sqrt{2}} \cdot \frac{1}{2}\right)$

$= \frac{(1+\sqrt{3})}{2\sqrt{2}}$

2. What is cos(15°)?

cos(15°) = cos(45° - 30°)

= cos(45°) * cos(30°) + sin(45°) * sin(30°)

$= \left(\frac{1}{\sqrt{2}} \cdot \frac{\sqrt{3}}{2}\right) + \left(\frac{1}{\sqrt{2}} \cdot \frac{1}{2}\right)$

$= \frac{(1+\sqrt{3})}{2\sqrt{2}}$

Alternatively, cos(15°) = cos(90° − 75°) = sin(75°)

3. Calculate tan(105°):

tan(105°) = tan(60° + 45°)

$= \frac{(\tan(60°)+\tan(45°))}{1-\tan(60°)\tan(45°)}$

$= \frac{(\sqrt{3}+1)}{1-\sqrt{3}}$

4. Calculate $\tan(135°)$:

$\tan(135°) = \tan(90°+45°)$

$\qquad = -\cot(45°)$

$\qquad = -1$

5. Determine $\cot(135°)$:

$\cot(135°) = 1 / \tan(135°)$

$= 1 / -1$

$= -1$

6. Find the value of $\sin(225°)$

$\sin(225°) = \sin(180°+45°)$

$\sin(225°) = -\sin(45°)$

$\sin(225°) = -\frac{1}{\sqrt{2}}$

7. Find the value of $\csc(225°)$:

$\csc(225°) = 1 / \sin(225°)$

$= \frac{1}{\left(-\frac{1}{\sqrt{2}}\right)}$

$= -\sqrt{2}$

8. What is sec(210°)?

sec(210°) = 1 / cos(210°)

$$= \frac{1}{\left(-\frac{\sqrt{3}}{2}\right)}$$

$$= \left(-\frac{2}{\sqrt{3}}\right)$$

$$= \left(-\frac{2\sqrt{3}}{3}\right)$$

9. sin (15°):

Step 1: Using half-angle formula: $\sin(15°) = \sqrt{\frac{(1-\cos(30))}{2}}$

Step 2: Substitute cos(30°) into the half-angle formula.

Final Value: $\sin(15°) = \frac{\sqrt{(2-\sqrt{3})}}{2}$

10. cos(105°):

Step 1: Using angle sum formula: cos(105°) = cos(45° + 60°)

cos(105°) = cos(45°)cos(60°) - sin(45°)sin(60°)

$$= \left(\frac{1}{\sqrt{2}} \cdot \frac{1}{2}\right) - \left(\frac{1}{\sqrt{2}} \cdot \frac{\sqrt{3}}{2}\right)$$

$$= \frac{(1-\sqrt{3})}{2\sqrt{2}}$$

GRAPHS OF BASIC TRIGONOMETRIC FUNCTIONS

$$y = \sin(x)$$

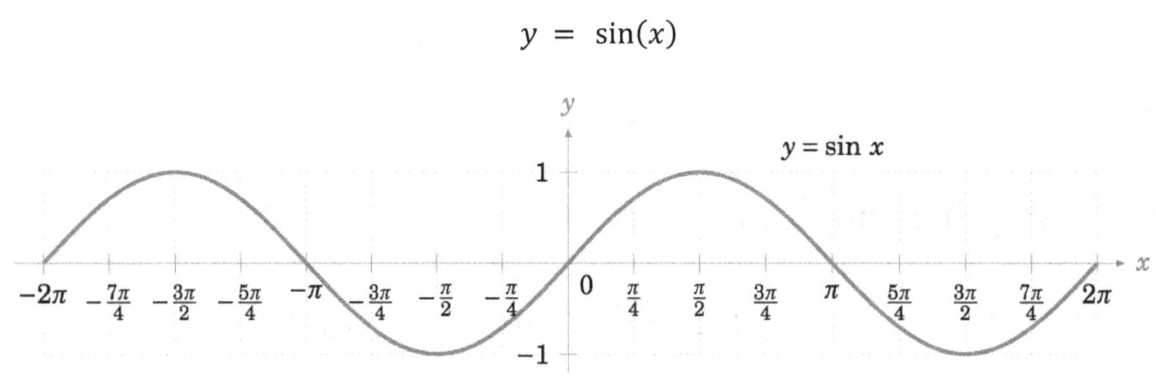

$y = \cos(x)$

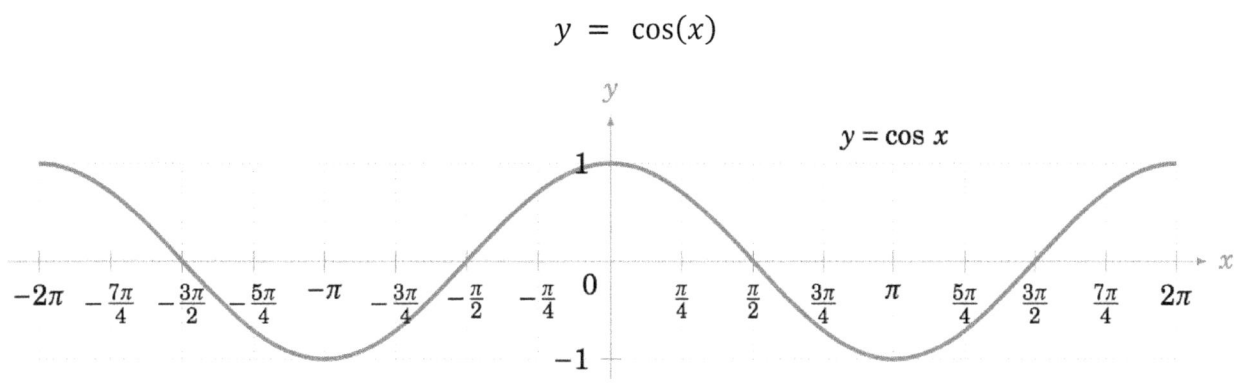

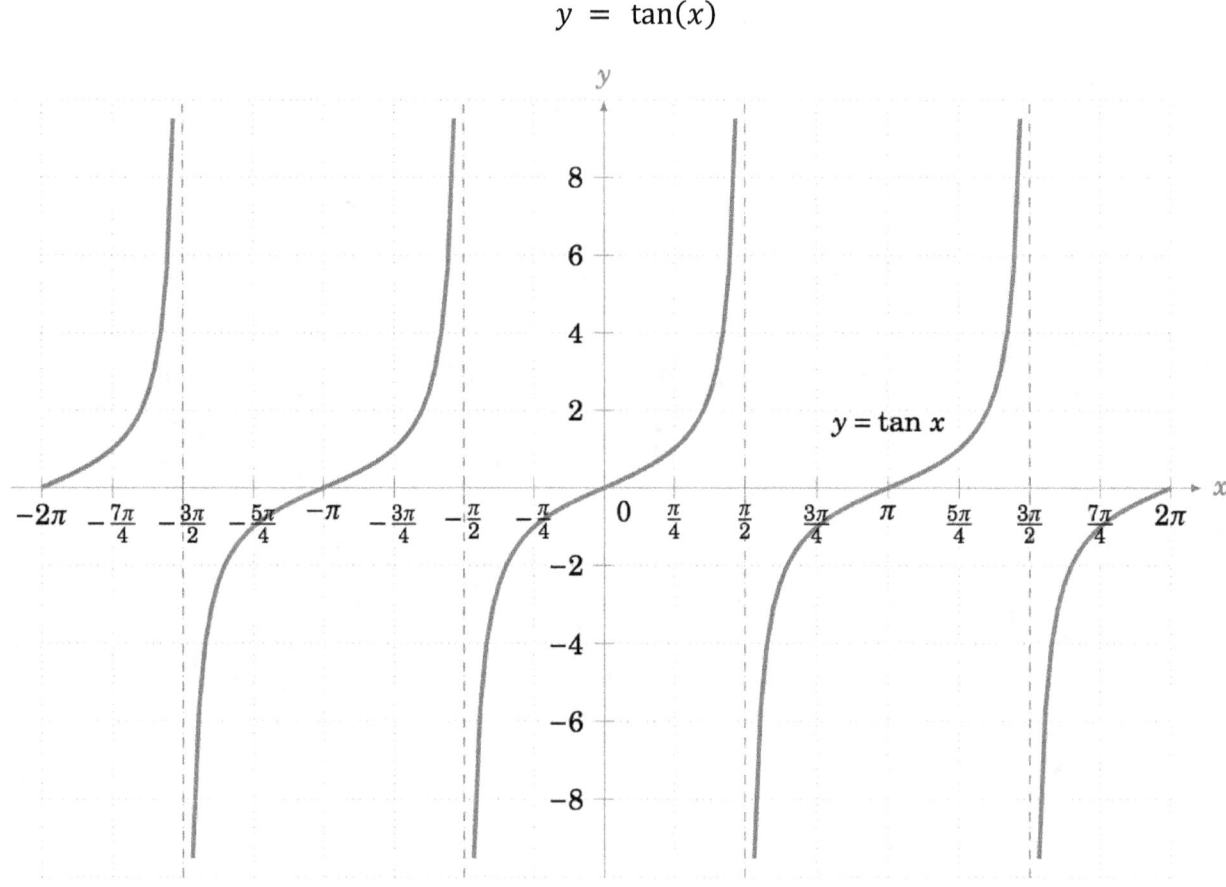

$y = \csc(x)$

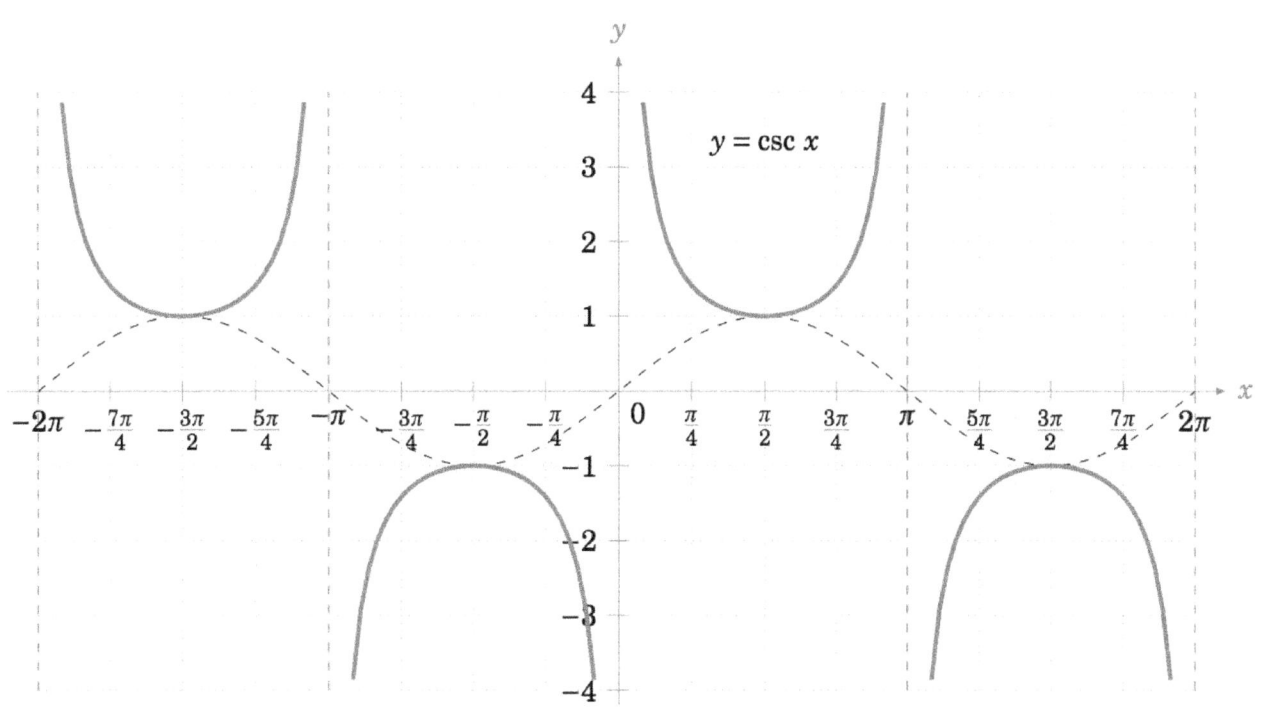

$$y = \sec(x)$$

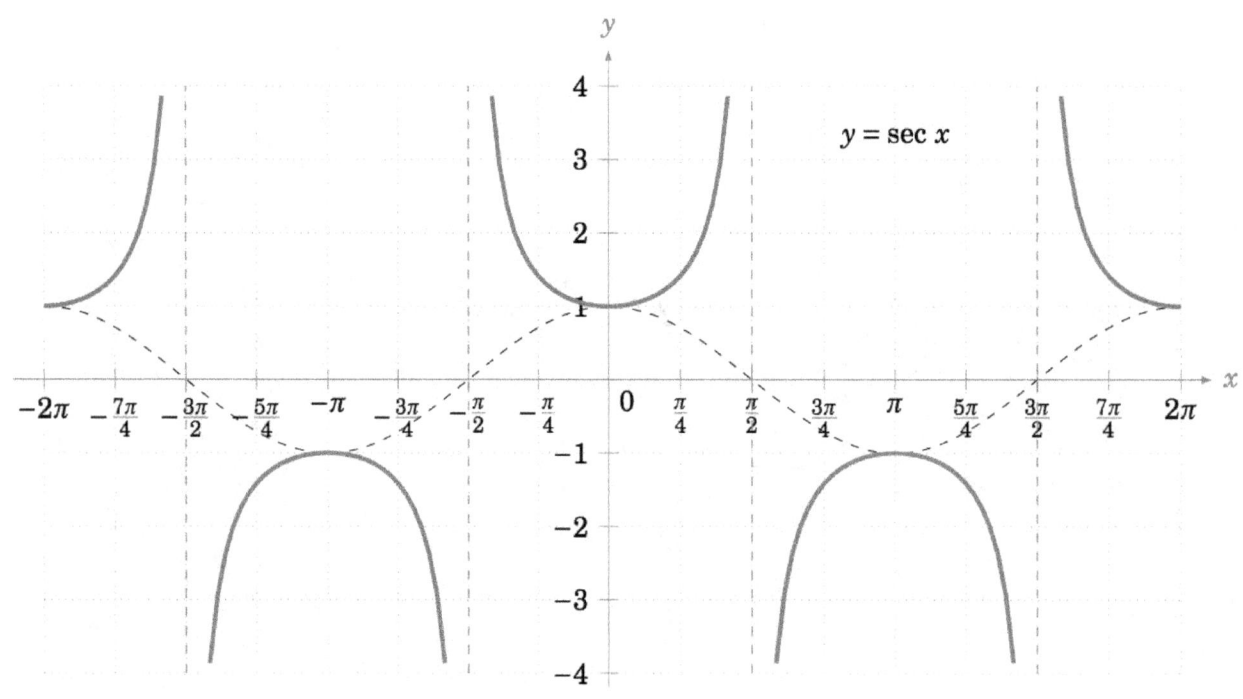

$y = \cot(x)$

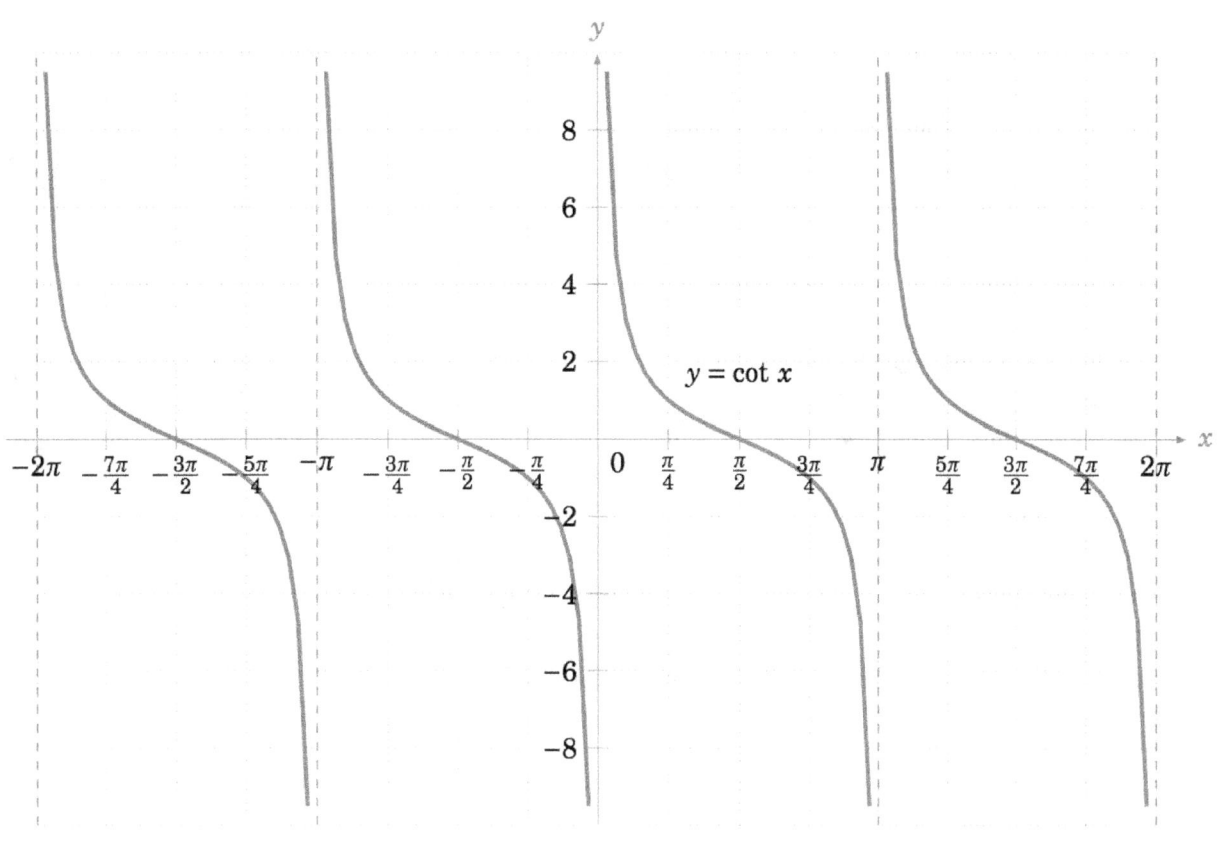

CHAPTER–4: HEIGHTS AND DISTANCES

IMPORTANT DEFINITIONS:

1. Line of Sight : It is the line drawn from the eye of an observer to the point in the object viewed by the observer.
2. Angle of elevation : The angle between the horizontal and the line of sight joining the observation point to an elevated object.
3. Angle of depression : The angle between the horizontal and the line of sight joining an observation point to an object below the horizontal level.

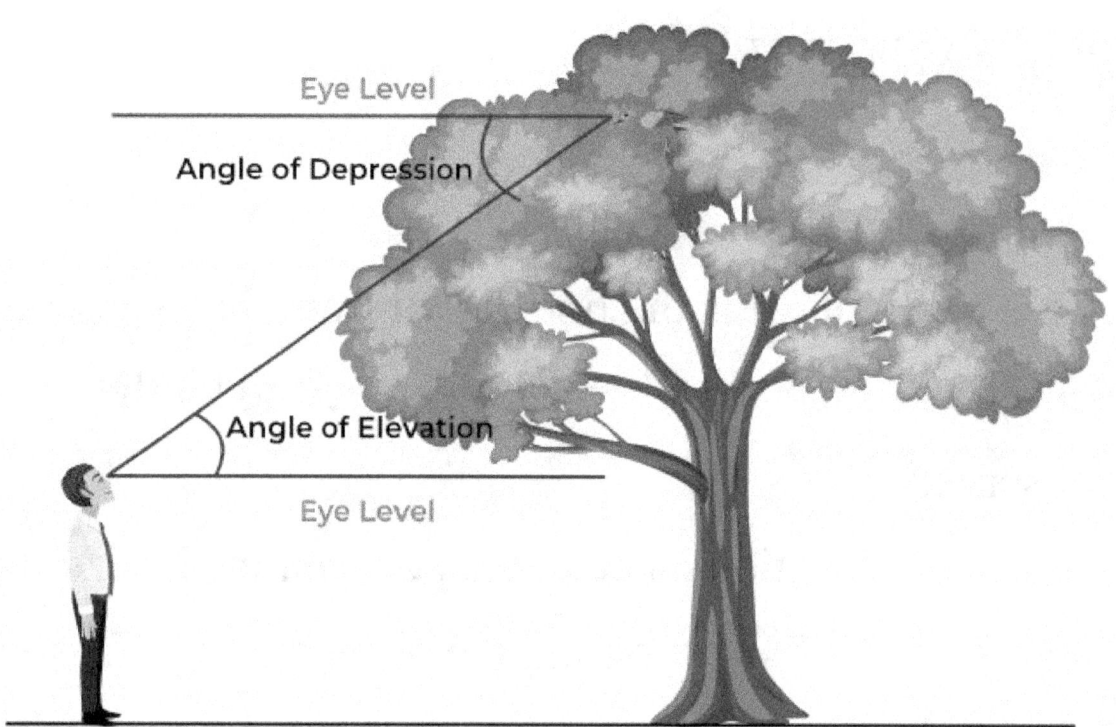

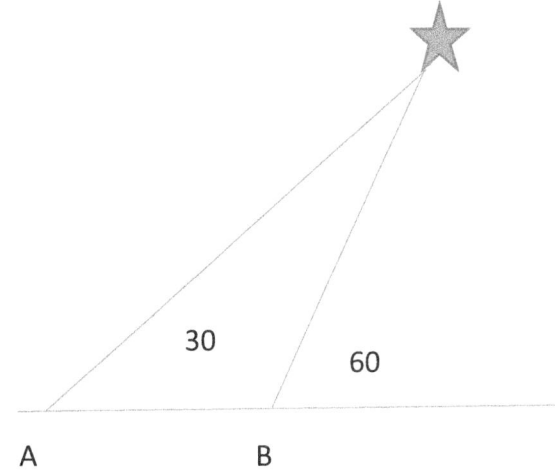

The angle of elevation of the star from point A is 30 degrees and from point B is 60 degrees.

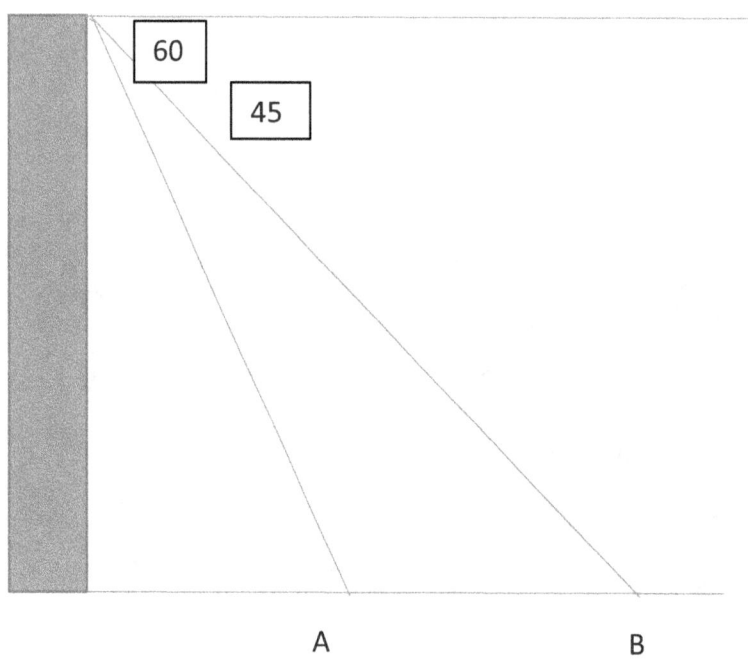

For example the observer is at the top of the building, then his angles of depression for the two view points namely A and B is 60 degrees and 45 degrees respectively.

Numericals:

Q1. A ladder reached the top of a vertical wall making an angle of 30 degrees with the ground, if the height of the wall is 12 metres, then find the length of the ladder.

Solution:

Let the length of the ladder be "l", then ;

$$\sin(30) = \frac{12}{l}$$

$$\frac{1}{2} = \frac{12}{l}$$

$$l = 24$$

Q2. A stay rod is used to keep the overhead poles straight by maintaining tension in opposite direction to the pull of line conductors. If the distance of the stay rod (connected from the pole to the ground) from the foot of the pole to point in the ground is X and if the stay rod makes an angle of θ with the ground, then what is the length of the rod in terms of X and θ ?

Solution: Let length of the rod be "l", then ;

$$\cos(\theta) = \frac{x}{l}$$

$$l = x\sec(\theta)$$

Q3. A man standing at a certain distance from a building, observe the angle of elevation of its top to be 60° . He walks 100 metres away from the building. Now, the angle of elevation of the building's top is 30° . How high is the building?

Solution:

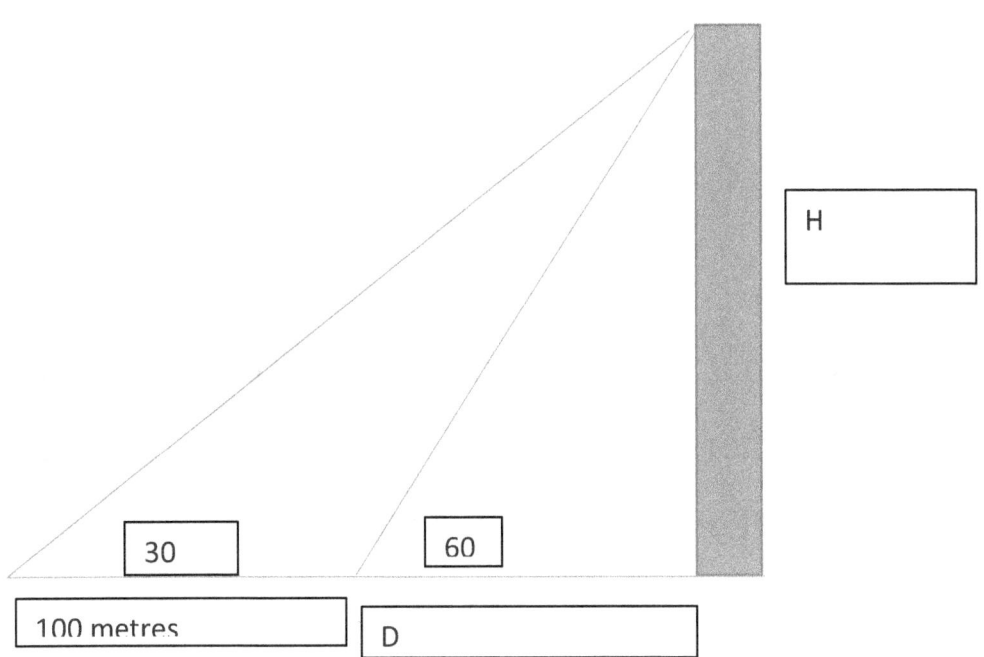

$$\tan(60) = \frac{H}{D}$$

$$H = D \cdot \sqrt{3}$$

$$\tan(30) = \frac{H}{D + 100}$$

$$H = \frac{(D + 100)}{\sqrt{3}}$$

$$\sqrt{3}D = \frac{(D+100)}{\sqrt{3}}$$

$$3D = D + 100$$

$$D = 50$$

$$H = 50.\sqrt{3} \quad \text{metres}$$

Q4. From a police tower on a road, the angle of depression of two cars approaching each other are α and β respectively. If the tower's height is h units, then find the distance between the two cars at the given angles.

Solution:

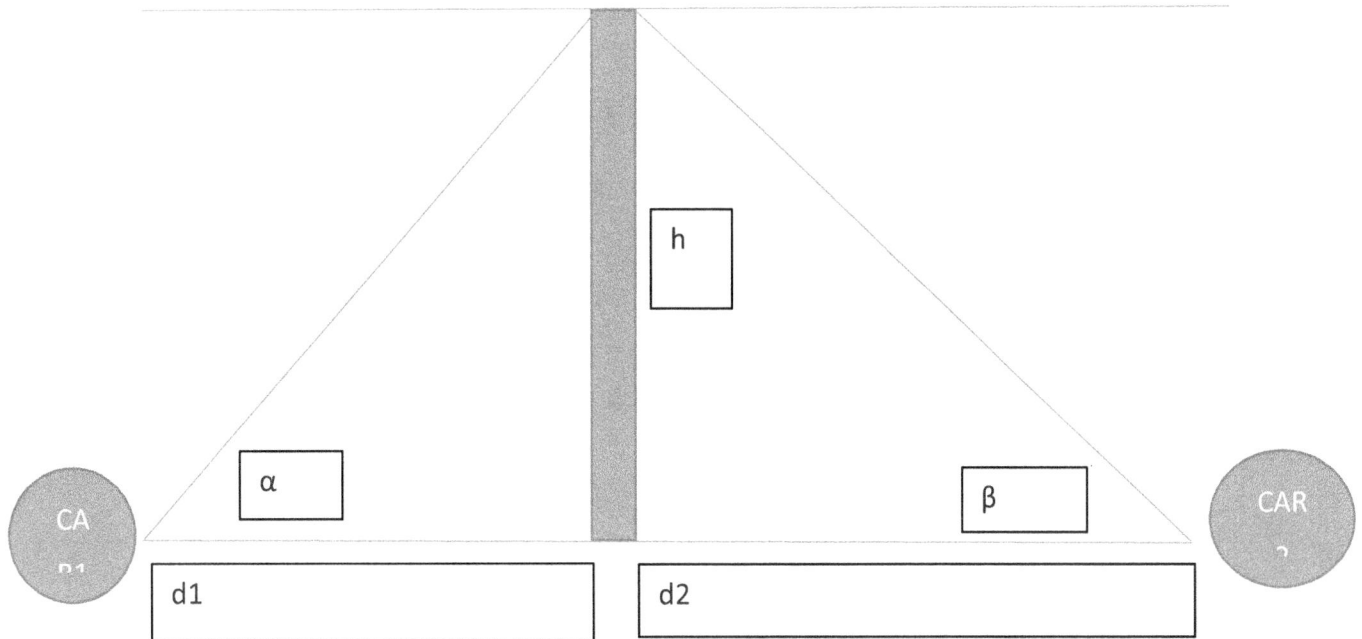

Let distance between the cars be D, then;

$$D = d1 + d2$$

$$\tan(\alpha) = \frac{h}{d1}$$

$$d1 = h \cdot \cot(\alpha)$$

$$\tan(\beta) = \frac{h}{d2}$$

$$d2 = h \cdot \cot(\beta)$$

$$D = h\cot(\alpha) + h\cot(\beta)$$

$$D = h[\cot(\alpha) + \cot(\beta)]$$

www.ingramcontent.com/pod-product-compliance
Lightning Source LLC
Chambersburg PA
CBHW082241220526
45479CB00005B/1298